I0091428

Diagnosing in Cardiovascular Chinese Medicine

Diagnosing in Cardiovascular Chinese Medicine

Dr. Anika Niambi Al-Shura, BSc., MSOM, Ph.D
Continuing Education Instructor
Niambi Wellness
Tampa, FL

Medical Illustrator: Samar Sobhy

AMSTERDAM • BOSTON • HEIDELBERG • LONDON
NEW YORK • OXFORD • PARIS • SAN DIEGO
SAN FRANCISCO • SINGAPORE • SYDNEY • TOKYO

ELSEVIER

Academic Press is an imprint of Elsevier

Academic Press is an imprint of Elsevier
32 Jamestown Road, London NW1 7BY, UK
The Boulevard, Langford Lane, Kidlington, Oxford, OX5 1GB, UK
Radarweg 29, PO Box 211, 1000 AE Amsterdam, The Netherlands
225 Wyman Street, Waltham, MA 02451, USA
525 B Street, Suite 1900, San Diego, CA 92101-4495, USA

Copyright © 2014 Elsevier Inc. All rights reserved.

No part of this publication may be reproduced or transmitted in any form or by any means,
electronic or mechanical, including photocopying, recording, or any information storage and
retrieval system, without permission in writing from the publisher. Details on how to seek
permission, further information about the Publisher's permissions policies and our arrangement
with organizations such as the Copyright Clearance Center and the Copyright Licensing Agency,
can be found at our website: www.elsevier.com/permissions

This book and the individual contributions contained in it are protected under copyright
by the Publisher (other than as may be noted herein).

Notices
Knowledge and best practice in this field are constantly changing. As new research and
experience broaden our understanding, changes in research methods, professional practices,
or medical treatment may become necessary.

Practitioners and researchers must always rely on their own experience and knowledge in
evaluating and using any information, methods, compounds, or experiments described herein.
In using such information or methods they should be mindful of their own safety and the safety
of others, including parties for whom they have a professional responsibility.

To the fullest extent of the law, neither the Publisher nor the authors, contributors, or editors,
assume any liability for any injury and/or damage to persons or property as a matter of products
liability, negligence or otherwise, or from any use or operation of any methods, products,
instructions, or ideas contained in the material herein.

British Library Cataloguing-in-Publication Data
A catalogue record for this book is available from the British Library

Library of Congress Cataloging-in-Publication Data
A catalog record for this book is available from the Library of Congress

ISBN: 978-0-12-800121-9

For information on all Academic Press publications
visit our website at **store.elsevier.com**

This book has been manufactured using Print On Demand technology. Each copy is produced
to order and is limited to black ink. The online version of this book will show color figures
where appropriate.

ELSEVIER | Book Aid International

**Working together
to grow libraries in
developing countries**

www.elsevier.com • www.bookaid.org

Transferred to Digital Printing in 2014

DEDICATION

The energy and effort behind the research and writing of this textbook is dedicated to my son, Khaleel Shakeer Ryland. May this inspire and guide you through your journey in your medical studies, career, and life.

ACKNOWLEDGMENTS

This is a special acknowledgement to my seven-year medical students at Tianjin Medical University (2012–2013) who served as cardiovascular research assistants. May your future medical careers be successful.

An Qi He
Bin Lin Da
Han Jiang
Chen Hua
Jia Ying Luo
Jun Zhang
Lin Lin
Ming Lu
Nang Zhang
Ping Tang
Hu Si Le
Zhao Tian Man
Wen Xing Ning
Xing Wen Zhao
Tang Ying Mei
Li Ying Ying
Xiong Yong Qin
Ding Yu
Li Yan Jun

CONTENTS

SECTION VI GAIT

NIAMBI WELLNESS
INTEGRATIVE CARDIOVASCULAR CHINESE MEDICINE

The companion course for this textbook edition can be found on the Elsevier website and at www.niambiwellness.com.

APPROVING AGENCIES

PROFESSIONAL
NCCAOM
DEVELOPMENT ACTIVITY ®

The course with this textbook is entitled, Integrative Diagnosing of Cardiovascular Diseases.

This course is approved by the National Certification Commission for Acupuncture and Oriental Medicine (NCCAOM), and is listed as course #1053-004 for 8 PDA points.

This course is approved by the Florida State Board of Acupuncture, and is listed as course # 20-334887 for 10 CEU credits.

COURSE DESCRIPTION

This course covers the finer details of cardiovascular patient examination techniques and clinical pearls used in Western medicine, together with TCM examination techniques.

COURSE OBJECTIVES

- Understand the six basic signs and symptoms connected to cardiac diseases according to Chinese medicine and Western medicine, and their clinical significance.
- Understand the methods and techniques used in physical examination to detect the signs.

Chest Pain

CHAPTER 1

Clinical Significance of Chest Pain

CHAPTER OBJECTIVES

After studying this chapter, you should be able to:

1. Explain the definition of chest pain according to Chinese medicine and Western medicine.
2. Describe the symptoms and differentiations according to Chinese medicine.
3. Describe the main Western medicine concerns during inquiry including possible other diagnosed diseases or disorders.

1.1 PART 1: DEFINITION

Chest pain is an unpleasant and possibly severe sensation, which originates or radiates around the chest. It is characterized by a feeling of pressure, tightness, or discomfort.

1. What is the diagnosis in Chinese medicine?

   _____

5. What are the contraindications of this formula?

   _____

   Visit the course to find the answers.

1.2 PART 2: ETIOLOGY

The pain signals may be transmitted by the vagus, and phrenic nerves, may also originate or be felt or palpated around the ribs and intercostal tissues or the sensation may be felt in the esophagus, diaphragm, scapulae, or spine.

2. What is the differential diagnoses in Western medicine?

Diagnosing in Cardiovascular Chinese Medicine. DOI: http://dx.doi.org/10.1016/B978-0-12-800121-9.00001-1
© 2014 Elsevier Inc. All rights reserved.

a. _____

b. _____

c. _____

d. _____

e. _____

5 What are the contraindications of this formula?

_____

Visit the course to find the answers.

It is important to determine the source of the pain:

Cardiovascular
Pulmonary
Other causes

Pain source	Causes	Symptoms
Cardiovascular	Ischemia	Myocardial infarction
	Non ischemic	Pericarditis or aneurysm
	Obstruction	Aortic disorders
Pulmonary causes	Chest wall	Pneumothorax and pleuritis
	Functional	Pneumonia, sarcoidosis, cancer, tuberculosis
	Vascular	Hypertension, embolus
	Cough	A sharp sound expelled from the lungs and throat and out of the mouth.
	Dyspnea	Difficult or labored breathing
Other causes	Psychological	Neurosis, depression, anxiety
	Gastroesophageal disorders	Acid reflux, nausea, dysphagia and vomiting

Copyright © 2014 Anika Niambi Al-Shura. Published by Elsevier Inc. All rights reserved.

Chinese medicine differentiates according to pulse, symptoms, signs, and at times the tongue may also give an important clue. Generally, the causes of chest pain will involve:

Stagnation
Deficiency

Chinese medicine causes and symptoms of chest pain		
Stagnation	Blood	Stabbing, fixed pain, sudden onset which is worse at night
	Phlegm	Chest oppression, heavy chest sensation, cough with mucous, fatigue and dizziness
	Cold	Pain which is worse with cold, shortness of breath, and a preference for warmth
Deficiency	Qi	Fatigue, low energy and shortness of breath
	Yin	Fatigue, night sweats, afternoon fever and malar flush
	Yang	Fatigue, cold body sensation, pale face, body edema and profuse clear urination

Copyright © 2014 Anika Niambi Al-Shura. Published by Elsevier Inc. All rights reserved.

4. Which cause is most common in the clinic?
   _____

5. What are the contraindications of this formula?
   _____

   Visit the course to find the answers.

NOTES

CHAPTER 2

Inquiry and Examination of Chest Pain in Western Medicine and TCM

CHAPTER OBJECTIVES

After studying this chapter, you should be able to:

1. Explain the reason for asking each initial question about the causes of chest pain.
2. Discuss the significance of auscultation for causes of chest pain.
3. Describe the cardiovascular causes of acute chest pain.
4. Describe the cardiovascular causes of chronic chest pain.

2.1 PART 1: INQUIRY AND EXAMINATION

Chest pain is the main reason patients visit a physician. The most important task is to determine whether the pain is likely caused by angina pectoris, myocardial ischemia, or coronary artery obstruction. In Chinese medicine, chest pain as well as the possible associated angina pectoris is classified under xiong bi chest obstruction and heartache.

7. What are the pathogenic factors?

   _____

8. What is the complication to avoid?

   _____

   _____

Visit the course to find the answers.

Diagnosing in Cardiovascular Chinese Medicine. DOI: http://dx.doi.org/10.1016/B978-0-12-800121-9.00002-3
© 2014 Elsevier Inc. All rights reserved.

Symptom	Pain quality in angina	Pain due to other causes
Pain	Sensation is crushing, dull or heavy	Stabbing, burning and sharp
Radiation	Involves the left, right arm or neck	Shoulders or back
Exertion	Symptoms are relieved by rest	May or may not be relieved by rest
Stress	No symptoms	Emotional or sexual events will trigger

Copyright © 2014 Anika Niambi Al-Shura. Published by Elsevier Inc. All rights reserved.

2.1.1 Inquiry

When interviewing the patient, the initial questions should include the following:

1. Describe the pain and symptoms.
2. Do you have any genetic variants which predispose you to atherosclerosis?
3. Do you have any other diagnosed diseases or disorders such as:
   9. _____
   10. _____
   11. _____

   Visit the course to find the answers.

Genetic factors:

DKN2A and DKN2B
These genes are involved with the formation of plaque within the arteries, which are a risk factor for coronary artery disease.
MTAP
This gene processes cellular waste products into methionine which metabolizes excess homocysteine.

12. What is the risk cardiovascular factor?

   _____

   Visit the course to find the answers.

2.1.2 Physical Examination

Measurement of temperature, waist circumference, weight, and height.

Evaluate the effects of blood pressure by viewing the retina, which can predict future patterns of cardiovascular disease.

Blood pressure:

- This is taken in sitting, standing, or supine positions and readings indicate stages of hypertension.

Pulses:

Chinese pulse diagnosis
- Irregular = weak heart qi
- Tight = pain and cold syndrome
- Weak = deficiency
- Wiry = qi stagnation
- Thready = hyperactive kidney
- Rolling = dampness, phlegm

Pulses		
Stagnation	Blood	Choppy, wiry or surging
	Phlegm	Rolling
	Cold	Tight, slow or deep
Deficiency	Qi	Rapid and weak
	Yin	Rapid and weak
	Yang	Slow and weak

Copyright © 2014 Anika Niambi Al-Shura. Published by Elsevier Inc. All rights reserved.

Western medicine:
- In Western medicine, the pulse may be considered bounding which is similar to choppy and surging.

Auscultation:

- First palpate the apical impulse for left ventricular enlargement or hypertrophy.
- Listen for an early sign of hypertension, the fourth heart sound (S4) which indicates left atrium overwork.
- Listen for underlying sign of heart disease, the third heart sound (S3) which indicates left ventricular malfunction.

Pathology:

- Regurgitation may be heard at the aortic position in aortic dissection.
- A pericardial friction rub may be heard in pericarditis.
- A mid-systolic click or late systolic murmur may be heard in mitral valve prolapse.

2.2 PART 2: ACUTE CHEST PAIN

Acute chest pain is characterized by an initial onset with a short or unstable duration. The main concern is whether the episode is due to angina pectoris (xiong bi) or myocardial infarction (possible separation of yin and yang). Calling for emergency assistance and transporting the patient to the hospital for a clear cause is essential.

13. What will an EKG show?

_____

Visit the course to find the answers.

Cardiovascular cause	Factors
Unstable Angina pectoris	The pain is relieved by rest or nitroglycerin.
	An electrocardiogram shows Q waves and ST segment depressions.
	Blood work includes CBC and cardiac enzyme test.
	Diagnosis can include coronary artery disease.
Myocardial infarction	The pain includes dyspnea or nausea and is not relieved by rest or nitroglycerin.
	An electrocardiogram shows Q waves and ST segment depressions.
	Blood work includes CBC and cardiac enzyme test.
Aortic dissection	The pain includes a ripping sensation in the chest, possible absent pulse and is not relieved by rest or nitroglycerin.
	Chest X-ray is important in diagnosing.
Pericarditis	The pain radiates to the left arm and is worse while lying down and relieved by sitting up.
	An electrocardiogram shows Q waves and ST segment depressions.

Copyright © 2014 Anika Niambi Al-Shura. Published by Elsevier Inc. All rights reserved.

14. What is the most significant test for unstable angina pectoris?

_____

15. What is the most significant test for myocardial infarction?

_____

16. What is the most significant test for aortic dissection?

_____

17. What is the most significant test for pericarditis?

_____

Visit the course to find the answers.

2.3 PART 3: CHRONIC CHEST PAIN

Chronic chest pain is characterized by an increase in intensity and/or frequency, with recurrent, predictable, and stable episodes. The main concerns include coronary artery disease and myocardial ischemia.

Cardiovascular cause	Factors
Coronary Artery Disease	The pain is relieved 3 min. after taking nitroglycerin.
Myocardial Ischemia	The pain is relieved after taking nitroglycerin and calcium channel blockers.
	An electrocardiogram shows Q waves and ST segment depressions.
	Blood work includes CBC and cardiac enzyme test.
Mitral Valve Prolapse	The pain is not relieved by rest or nitroglycerin.
	Doppler and EKG is important in diagnosing, especially for vegetations in endocarditis and possible arrythmias.
Pericarditis	The pain radiates to the left arm and is worse while lying down and relieved by sitting up.
	An electrocardiogram shows Q waves and ST segment depressions.

Copyright © 2014 Anika Niambi Al-Shura. Published by Elsevier Inc. All rights reserved.

18. What is most significant about myocardial infarction?
_____

19. What is most significant about mitral valve prolapse?
_____

20. What is most significant about pericarditis?
_____

Visit the course to find the answers.

Other causes	Factors
Esophageal Disorders	The pain happens due to movement of stomach acids: lying down or eating.
	Spasms are detected along the walls of the esophagus.
Mitral Valve Prolapse	The pain is not relieved by rest or nitroglycerin.
	Doppler and EKG is important in diagnosing, especially for identifying vegetation in endocarditis and possible arrhythmias.
Pericarditis	The pain radiates to the left arm and is worse while lying down and relieved by sitting up.
	An electrocardiogram shows Q waves and ST segment depressions.

Copyright © 2014 Anika Niambi Al-Shura. Published by Elsevier Inc. All rights reserved.

NOTES

Log on at www.niambiwellness.com to access the companion course and quiz for Module 1.

Breathing Difficulties

CHAPTER 3

Clinical Significance of Dyspnea and Orthopnea

CHAPTER OBJECTIVES

After studying this chapter, you should be able to:

1. Explain the definition of dyspnea and orthopnea.
2. Describe the mechanisms which contribute to rapid, shallow breathing.
3. Describe the complications which contribute to decreased pulmonary compliance.

3.1 PART 1: DEFINITION

Dyspnea is an uncomfortable sensation in the chest which is difficult to control while breathing.

1. What does the sensation suggest?

   _____

   Visit the course to find the answers.

   Orthopnea is an uncomfortable sensation of breathlessness which wakes a person while sleeping or is difficult to control while lying down.

2. What can relieve the sensation?

   _____

   Visit the course to find the answers.

   In Chinese medicine, both breathing problems are characterized by lung qi deficiency with a failure of the kidney to grasp the qi. Damp phlegm obstruction causes asthma.

3. When is this condition most commonly triggered?

   _____

   Visit the course to find the answers.

Diagnosing in Cardiovascular Chinese Medicine. DOI: http://dx.doi.org/10.1016/B978-0-12-800121-9.00003-5
© 2014 Elsevier Inc. All rights reserved.

3.2 PART 2: ETIOLOGY

3.2.1 Causes of Dyspnea and Orthopnea

Cardiovascular diseases often cycle with respiratory disorders and interfere with activity tolerance and breathing at rest.

- A decrease in lung ventilation capacity, such as with pleural effusion in the interstitial spaces around the alveoli, causes stimulation of the J-receptors of the vagus nerve. The Hering–Breuer reflex is then activated causing rapid, shallow breathing.
- An acid–base imbalance, hypoxia, anemia, and oxygen deficit.
- Complications such as asthma, chronic obstructive pulmonary disease, and congestive heart failure can cause airway resistance due to decrease in bronchial flow or systemic edema.

NOTES

CHAPTER 4

Inquiry and Examination of Dyspnea and Orthopnea in Western Medicine and TCM

CHAPTER OBJECTIVES

After studying this chapter, you should be able to:

1. Explain a Chinese medicine perspective of the cause of breathlessness.
2. Discuss the significance of auscultation for diagnosing dyspnea.
3. Describe the cardiovascular significance of dyspnea.
4. Describe the cardiovascular significance of orthopnea.

4.1 PART 1: INQUIRY AND EXAMINATION

4.1.1 Inquiry

When you ask the patient a question about their breathing, you should begin with a nonleading question.

Q: Have you experienced any difficulty in breathing?

When you ask the patient a question about personal history, you should inquire for details:

4. _____
5. _____
6. _____
   Visit the course to find the answers.

When you ask the patient a question about the onset, you should inquire for details:

7. _____
8. _____
9. _____
   Visit the course to find the answers.

When you ask the patient a question about the quality or other symptoms, you should inquire for details:

10. _____
11. _____

Diagnosing in Cardiovascular Chinese Medicine. DOI: http://dx.doi.org/10.1016/B978-0-12-800121-9.00004-7
© 2014 Elsevier Inc. All rights reserved.

12. _____
13. _____
14. _____

 Visit the course to find the answers.

When you ask the patient a question about personal ability or limitations, you should inquire for details:

15. _____
16. _____

 Visit the course to find the answers.

When you ask the patient a question about sleeping quality and duration, you should inquire for details:

17. _____
18. _____
19. _____

 Visit the course to find the answers.

 What is done to relieve the symptoms?

4.1.2 Physical Examination

While inquiring the patient about habits and symptoms, one Chinese medicine perspective will classify them into the category of breathlessness with the differentiation of qi deficiency. The patient reporting the ability to exhale, but difficulty inhaling most likely has a failure of the kidneys to grasp the qi. The tongue and pulse examination may or may not be a significant diagnostic method to determine diagnosis, especially when the commonly noticed signs will include a pale tongue and a weak and empty pulse.

20. Why is it necessary to inspect the pulse?

 _____

 Visit the course to find the answers.

Blood pressure:
- This is taken in sitting, standing, or supine positions and readings indicate stages of hypertension

Pulses:
Common Chinese pulse diagnosis:
- Weak or absent = qi deficiency
- Thready = hyperactive kidney
- Rolling = dampness, phlegm

Pulses	
Lung qi	Weak
Failure of kidney to grasp qi	Deep and weak

Copyright © 2014 Anika Niambi Al-Shura. Published by Elsevier Inc. All rights reserved.

A. Compare the Chinese medicine pulses with the Western medicine pulse.

Western medicine:

- Absent

Auscultation:

- First palpate the apical impulse for left ventricular enlargement or hypertrophy.
- Listen for an early sign of hypertension, the fourth heart sound (S4) which indicates left atrium overwork.
- Listen for underlying sign of heart disease, the third heart sound (S3) which indicates left ventricular malfunction.

Pathology:

- Regurgitation may be heard at the aortic position in aortic dissection.
- A pericardial friction rub may be heard in pericarditis.
- A mid-systolic click or late systolic murmur may be heard in mitral valve prolapse.

4.2 PART 2: DYSPNEA

4.2.1 Dyspnea on Exertion

Dyspnea is difficult or uncomfortable breathing during exertion. Following regular movement, the patient will also complain of fatigue. Dyspnea is a symptom seen in patients with atrial fibrillation, pericardial effusion, cardiomyopathy, myocarditis, mitral stenosis, aortic insufficiency, renal failure, after myocardial infarction, and left ventricular failure. In left ventricular failure.

21. What is the cause of dyspnea?
   _____

   Visit the course to find the answers.

4.2.2 Nocturnal Dyspnea

Paroxysmal nocturnal dyspnea is a condition in patients with left and right ventricular heart failure and increased pulmonary fluid pressure. The patient is suddenly awakened while sleeping in a prone or supine position.

22. What provides relief?

_____

Visit the course to find the answers.

4.3 PART 3: ORTHOPNEA

Orthopnea awakens the prone or supine lying patient, anytime while sleeping. The patient must sit or stand for relief. In these patients it is likely they already have congestive heart failure. There is failure in both ventricles and increased pressure of fluid through the pulmonary circulation.

23. When does the pressure decrease?

_____

Visit the course to find the answers.

NOTES

Log on at www.niambiwellness.com to access the companion course and quiz for Module 2.

Heart Rhythm

Heart Rhythm

CHAPTER 5

The Clinical Significance of Palpitations

CHAPTER OBJECTIVES

After studying this chapter, you should be able to:

1. Explain the definition of palpitations in Western medicine and Chinese medicine.
2. Discuss the etiology of palpitations.
3. Describe the noncardiac and cardiac causes of palpitations.

5.1 PART 1: DEFINITION

Heart palpitations are unpleasant awareness of prominent heartbeats. They may be characterized as a sensation of the heart pounding and fluttering, racing very fast, and skipping beats as a result. Arrhythmia is categorized in Chinese medicine as

1. _____
2. _____
3. _____

Visit the course to find the answers.

5.2 PART 2: ETIOLOGY

The presence of palpitations reflects regular yet abnormal adjustments in rate, rhythm, and contractility. In most cases, an extra systolic beat may occur with a premature beat. Without accompanying symptoms, it is inconclusive to determine that palpitations are considered significant enough to indicate the presence of a heart disease. In heart disease, factors can include

Diagnosing in Cardiovascular Chinese Medicine. DOI: http://dx.doi.org/10.1016/B978-0-12-800121-9.00005-9
© 2014 Elsevier Inc. All rights reserved.

4. _____
5. _____
6. _____
7. _____

Visit the course to find the answers.

Palpitation causes	Description
Non cardiac	Some non cardiac symptoms which occur in conjunction with palpitations include anxiety and depression. These symptoms are also part of a sedentary lifestyle.
Cardiac	Palpitations are a major feature of arrhythmias and can be experienced at rest, while sleeping, and during the process of moving from a sitting to a standing position in orthostatic hypotension. Palpitations can also be connected to serious stages of hypertension, and anemia which may affect cardiac output.

Copyright © 2014 Anika Niambi Al-Shura. Published by Elsevier Inc. All rights reserved.

8. What two non-cardiac causes most commonly are associated with palpitations?

_____

_____

Visit the course to find the answers.
Visit the course to find the answers.

9. What three activities may feature the cardiac causes of palpitations?

_____

_____

_____

The health practitioner must review the complete patient history and determine whether there are other cardiovascular symptoms such as chest pain and dyspnea or existing heart diseases such as ischemia or myopathy.

10. What clues can the patient history provide?

_____

Visit the course to find the answers.

Type of arrhythmia	Quality of rhythm sensation
Paroxysmal Atrial fibrillation	The rhythms are irregular
Normal Sinus Tachycardia	The rhythms are irregular
Sinus Tachycardia	The heart rhythm increases abruptly in a regular rapid rhythm.
Paroxysmal Supraventricular	The heart rhythm increases abruptly then terminates into regular palpitations.

Copyright © 2014 Anika Niambi Al-Shura. Published by Elsevier Inc. All rights reserved.

11. What are the similarities and differences between the types of arrhythmia?

S:_____

D:_____

Other serious factors related to palpitations:

12. _____
13. _____
14. _____
15. _____
Visit the course to find the answers.

NOTES

CHAPTER 6

Inquiry and Examination of Palpitations in Western Medicine and TCM

CHAPTER OBJECTIVES

After studying this chapter, you should be able to:

1. Explain a Chinese medicine perspective of the cause of palpitations.
2. Discuss the significance of genetic factors.
3. Describe the Chinese and Western medicine integrated differentiations.
4. Describe the appearance of the EKG waves associated with certain arrhythmias.

6.1 PART 1: INQUIRY

Palpitations are another major reason why patients visit a physician. In some cases patients may not have the opportunity to make emergency medical care in time due to sudden death.

In Chinese medicine, palpitations are characterized as xiong bi or chest obstruction, xin ji or palpitation, and xuan yun or vertigo. Tachycardia and bradycardia are characteristics of the beat pattern when investigating palpitations. Tachycardia is often differentiated as blood stagnation and qi and yin deficiency. Bradycardia is usually differentiated as blood stasis which blocks the heart vessel, phlegm, yang deficiency, cold, and dampness. Health practitioners are interested in quality of the heartbeats which can be considered an irregularity or compensation.

16. What does the history and diary help determine?
   _____

   Visit the course to find the answers.

Diagnosing in Cardiovascular Chinese Medicine. DOI: http://dx.doi.org/10.1016/B978-0-12-800121-9.00006-0
© 2014 Elsevier Inc. All rights reserved.

6.1.1 Inquiry

When you ask the patient a question about the sensation of the unpleasant heartbeat, you should ask about the quality of the beats:

17. _____

Visit the course to find the answers.

When you ask the patient a question about the sensation of the unpleasant heartbeat, you should ask about the rhythm of the beats:

18. _____

Visit the course to find the answers.

When you ask the patient a question about the sensation of the unpleasant heartbeat, you should ask about the history related to the beats:

19. _____
20. _____
21. _____
22. _____
23. _____
24. _____
25. _____
26. _____
27. _____

Visit the course to find the answers.

6.1.2 Patient Symptoms

When you ask patients about any history of dizziness or vertigo, listen for special sensations related to dizziness or the floor rising up.

When you determine the Chinese medicine cause tachycardia or bradycardia consider the feelings associated with each type of arrhythmia.

Genetic factors:

Genetic factors assist in determining the nature and magnitude of the disease or symptoms and the likelihood of success in treatment.

KCN (potassium channel) gene mutations and Familial Atrial Fibrillation.

Three mutations on genes KCNE2, KCNJ2, and KCNQ1 involve the replacement of certain key amino acid proteins which make up the channels that regulate potassium flow which assists regular heart rhythm. The replacement or alterations interfere with channel reception, affecting ion supply, and exchange during a relevant phase, with often detrimental outcomes.

- Romano-Ward syndrome/long QT syndrome:
  Potassium channels open slower and close faster than usual. The decreased outflow of potassium ions causes syncope and possible sudden death.
- Short QT syndrome:
  The increased flow of potassium out of cells causes the irregular shortened interval between heartbeats, which may cause syncope and sudden death.
- Andersen−Tawil syndrome:
  This condition causes arrhythmia marked by muscle weakness and paralysis due to mutations which change the shape of the channel causing potassium transport problems.
- Jervell and Lange-Nielsen syndrome:
  The KCNQ1 has fewer, shorter proteins and unable to help build potassium channels which causes congenital arrhythmia and deafness.
- Long QT syndrome:
  Potassium channels are affected causing polarization failure between the action and the resting potential phases.

6.1.3 KCNE2 Gene
The amino acid proteins of this gene form the structure of the potassium voltage-gated channel, which transports potassium in and outside of cardiac myosites, and maintain the normal rhythm. Cardiovascular diseases are involved in a mutation of this gene:

- Romano-Ward syndrome/long QT syndrome.

6.1.4 KCNJ2 Gene
This channel is called the potassium inwardly rectifying channel and is responsible for generation of signals in the conduction system. These channels are found in cardiac and muscle tissue and assist with

contraction and relaxation. Cardiovascular diseases are involved in a mutation of this gene:

• Short QT syndrome
• Andersen−Tawil syndrome

6.1.5 KCNQ1 Gene

This gene and the KCNE1 are part of the components of potassium channels. Channels with this protein regulate cardiac rhythm, found in the ear for balance and hearing, and for functioning of lung, kidney, and gastrointestine. In addition, this gene has been identified in sudden death syndrome and also triggered during menopause with diseases such as ovarian cancer, type II diabetes, and long QT syndrome. Cardiovascular diseases are involved in a mutation of this gene:

• Jervell and Lange-Nielsen syndrome
• Romano-Ward syndrome/long QT syndrome
• Short QT syndrome
• Long QT syndrome

6.1.6 ACE Gene Polymorphism

Patients with this gene may experience altered responses to ACE inhibitor medication. This is significant when patients are considered for herbal formula therapy rather than pharmaceutical drugs.

6.2 PART 2: PHYSICAL EXAMINATIONS

The clinical examination of palpitations should include

28. _____
29. _____
    Visit the course to find the answers.

In addition, an EKG recording using a holter monitor should be done during an episode.

While inquiring the patient about habits and symptoms, one Chinese medicine perspective will automatically classify them into the category of xiong bi with the differentiation of qi deficiency. The tongue and pulse examination may or may not be a significant method to determine diagnosis, especially when the commonly noticed signs may include a pale tongue and a weak and empty pulse. However, it is still necessary to inspect, as the pulse may change when the condition includes an exterior syndrome or a cardiovascular complication.

Blood pressure:
- This is taken in sitting, standing, or supine positions and readings indicate stages of hypertension as a possible factor.

Pulses:
- Common Chinese pulse diagnosis:
- Irregular = weak heart qi
- Tight = pain and cold syndrome
- Weak = deficiency
- Wiry = qi stagnation
- Thready = hyperactive kidney
- Rolling = dampness, phlegm

Western medicine:
- In Western medicine, the irregular or intermittent pulse which "skips beats" is used in diagnosing palpitations.

Auscultation:
- First palpate the apical impulse for left ventricular enlargement or hypertrophy.
- Listen for an early sign of hypertension, the fourth heart sound (S4) which indicates left atrium overwork.
- Listen for underlying sign of heart disease, the third heart sound (S3) which indicates left ventricular malfunction.

Pathology (see Table 1.9).

6.3 PART 3: CARDIAC CONCERNS

With the presence of disease, palpitations may be a positive sign.

If it is cardiac, then what should you notice on EKG?

30. _____

Visit the course to find the answers.

6.4 PART 4: VASCULAR CONCERNS

Palpations may include adverse changes in the vascular system, which can be seen in the blood pressure.

Blood pressure reading may reflect signs such as hypertension or hypotension.

6.5 PART 5: AUSCULTATION

There are two common types of tachycardia that are noticed during examination:

1. Paroxysmal atrial fibrillation
2. Sinus tachycardia

6.5.1 Paroxysmal Atrial Fibrillation

These irregularities in rhythm or "skipping" of beats are seen as disorganized activity and a missing P wave on EKG. This pattern is sometimes considered a normal normal rhythm. Increased intensity or "racing" of the beats may indicate ventricular or supra-ventricular tachycardia. This could indicate a serious heart condition which could result in sudden death. Specialist attention in Western medicine may be needed. Emergency assistance is required if patient begins to present with signs of distress ie: clutching chest, breathlessness, syncope, etc. Other findings may reveal:

31. _____
32. _____
33. _____
34. _____
Visit the course to find the answers.

6.5.2 Sinus Tachycardia

The sensation is a gradual onset with the heart rate at or above 100 bpm. On EKG, the P and T waves are seen close together yet is sometimes considered a normal normal rhythm. Types of A−V heart block may be seen on EKG in seriously ill patients. The causes are broad and include the following:

35. _____
36. _____
37. _____
38. _____
Visit the course to find the answers.

6.6 PART 6: EKG CONFIRMATIONS

Arrhythmias:
a. Ventricular tachycardia

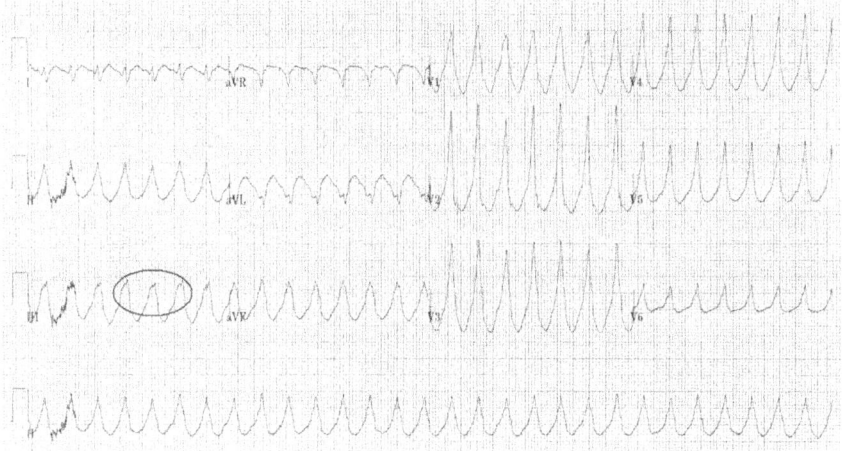

Copyright © 2014 Anika Niambi Al-Shura. Published by Elsevier Inc. All rights reserved.

b. Paroxysmal atrial fibrillation

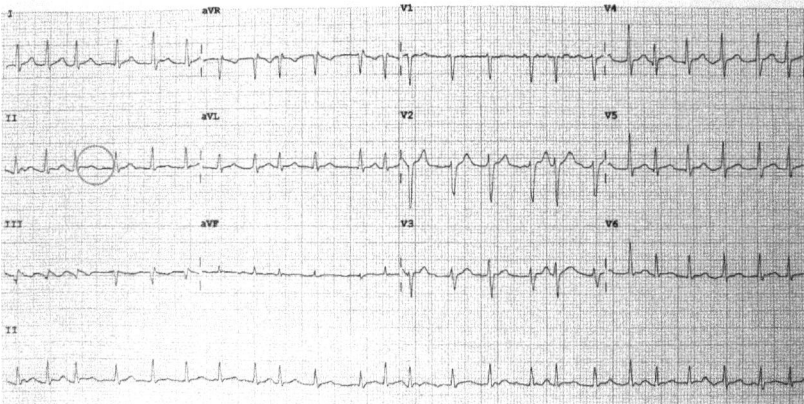

Copyright © 2014 Anika Niambi Al-Shura. Published by Elsevier Inc. All rights reserved.

c. Sinus tachycardia

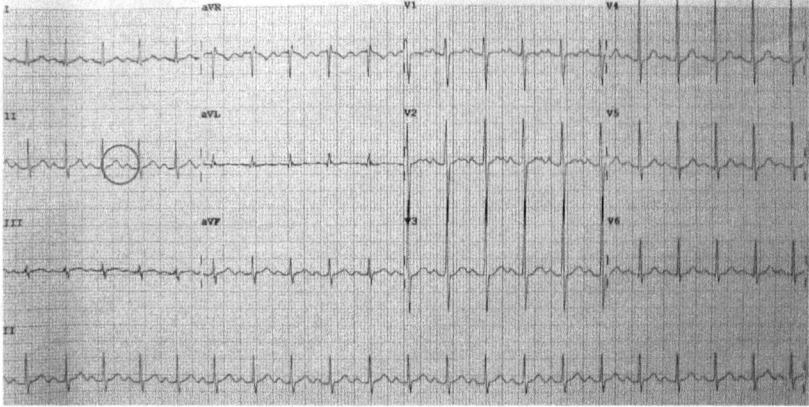

Copyright © 2014 Anika Niambi Al-Shura. Published by Elsevier Inc. All rights reserved.

d. Bundle branch block (atria)

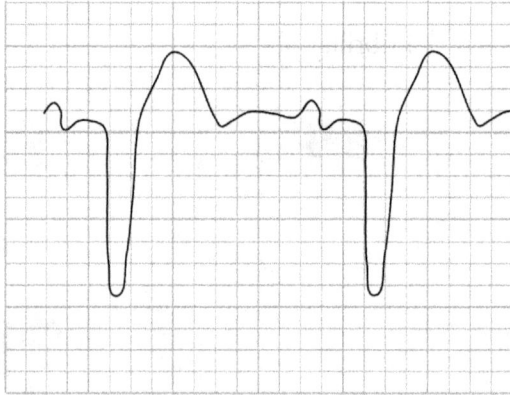

Copyright © 2014 Anika Niambi Al-Shura. Published by Elsevier Inc. All rights reserved.

NOTES

Log on at www.niambiwellness.com to access the companion course and quiz for Module 3.

Fainting Spells

CHAPTER 7

Clinical Significance of Syncope

CHAPTER OBJECTIVES

After studying this chapter, you should be able to:

1. Explain the definition of syncope in Western medicine and Chinese medicine.
2. Discuss the etiology of syncope.
3. Discuss how cardiac output and cerebral blood flow is involved in syncope.

7.1 PART 1: DEFINITION

Syncope is defined as a sudden loss of consciousness due to a sudden decrease in blood pressure followed by decrease in cerebral blood flow. In Chinese medicine, syncope may be considered.

1. _____
2. _____
   Visit the course to find the answers.

7.2 PART 2: ETIOLOGY

The mechanisms of syncope are varied and complicated. Syncope can also be associated with arrhythmias and may have unknown causes as well. The main physiological factors involve impulses from baroreception which acts in vasodilation and constriction, and areas of the neurological system such as the vagus nerve, glossopharyngeal nerve which affects the function of the sinus node. Some other types may be under the category of vasodepressor syncope, which is basically fainting in different events such as having blood drawn, grief, emotional, or psychological situations. Micturition syncope occurs with relieving of

Diagnosing in Cardiovascular Chinese Medicine. DOI: http://dx.doi.org/10.1016/B978-0-12-800121-9.00007-2
© 2014 Elsevier Inc. All rights reserved.

urine. Defecation syncope occurs with relieving of feces. In Chinese medicine, yuan xuan is also in the category of yun jue which involves what factors that contribute to syncope?

3. _____

4. _____

5. _____

Visit the course to find the answers.

The two main types discussed in this chapter are related to cardiac output and cerebral blood flow.

7.2.1 Cardiac Output

Syncope is often a factor related to arrhythmias. Several factors point to ventricular factors associated with outflow failures. Right ventricular outflow failure contributes to pulmonary hypertension. Insufficient cardiac output associated with left ventricular outflow failure, leads to decreased cerebral blood flow.

7.2.2 Cerebral Blood Flow

Problems with the cerebral arteries such as in ischemia or emboli may lead to syncope.

The reticular activating system is made up of the brainstem and cortex and controls consciousness as in being awake and alert. It is most affected by ischemia.

The vertebrobasilar arteries are most affected by emboli. Both disorders may contribute to syncope and also neurological disorders such as stroke and can also lead to death.

NOTES

Log on at www.niambiwellness.com to access the companion course for Chapter 1 review and quiz.

CHAPTER 8

Inquiry and Examination of Syncope in Western Medicine and TCM

CHAPTER OBJECTIVES

After studying this chapter, you should be able to:

1. Explain a Chinese medicine perspective of syncope.
2. Discuss the significance of genetic factors.
3. Describe the role of arrhythmias and structural problems in syncope.
4. Describe the inquiry process and how answers might distinguish the cardiac from the noncardiac causes.

In Chinese medicine, syncope may be considered yuan xuan and in the category of yun jue, which involves blood stasis with qi and blood deficiency and phlegm. The matter of diagnosing, treating, and monitoring patients with well-documented cases of syncope can be a complicated process. Most patient diagnoses are often initiated in the Western medicine clinical setting with monitored referrals for Chinese medicine evaluation and treatment.

8.1 PART 1: INQUIRY

When you ask the patient a question about episodes of syncope, you should inquire about whether there was any accompanying factors which occurred prior such as chest pain, dyspnea, palpitations, pain located in the head or back, memory problems, vision problems, speech problems, or loss of postural tone. Additional diagnostic testing can help determine if the syncope is cardiac or cerebral.

Diagnosing in Cardiovascular Chinese Medicine. DOI: http://dx.doi.org/10.1016/B978-0-12-800121-9.00008-4
© 2014 Elsevier Inc. All rights reserved.

Begin with an inquiry about frequency:

6. _____

Visit the course to find the answers.

When you ask the patient a question about personal history, you should inquire for details:

7. _____
8. _____

Visit the course to find the answers.

When you ask the patient a question about the personal issues which occur around episodes of syncope, you should inquire for these details:

9. _____
10. _____

Visit the course to find the answers.

8.2 PART 2: PHYSICAL EXAMINATIONS

The main causes of syncope discussed in this chapter involve cardiac output and cerebral flow; however, there are other unknown causes to consider. In some patients, a physical examination and other evaluations may not provide the origins. The possible reasons for this may be:

11. _____
12. _____

Visit the course to find the answers.

These patients may need to be referred to a specialist who can monitor and provide additional diagnostic testing.

8.2.1 Noncardiac Causes

Type of Syncope	Description
Vasodepressor Syncope	This form of syncope happens in response to emotional stress concerning a real injury or a threat. It occurs when the patient rises from lying or sitting to the standing position.
Micturition Syncope	This form of syncope usually occurs after urination, and may be due to malnourishment, exhaustion, infection, excessive illicit drug use, alcohol intake, medications such as diuretics and orthostatic hypotension.
Defecation Syncope	This form of syncope usually occurs in the elderly and may be of unknown cause.
Orthostatic Hypotension	This form of syncope occurs when the blood pressure drops to< 90 mmHg. It may be due to central and peripheral nervous system disorders, blood volume depletion leading to decreased venous return, and certain medications.
Swallow Syncope	This form of syncope is rare but can be connected with bradycardia and AV block. It occurs with esophageal diseases.
Cough Syncope	This form of syncope occurs in patients who are heavy smokers with cardiomyopathies and pulmonary obstructive disease.
Carotid sinus Syncope	This form of syncope occurs in elderly patients who wear tight shirt collars. Episodes are caused by head or neck movement which stretches the carotid sinus. Medications such as digoxin and propanolol, and also tumors, and inflammation may cause occurrence.
Exertional Syncope	This form of syncope occurs in patients during exertion of brute effort, while playing sports and in some cases during regular movement.

Copyright © 2014 Anika Niambi Al-Shura. Published by Elsevier Inc. All rights reserved.

8.2.2 Patient Symptoms
Blood pressure:

13. _____

14. _____

Visit the course to find the answers.

Pulses:

Common Chinese pulse diagnosis:

Pulses	
Blood stagnation	Choppy
Blood deficiency	Weak
Qi deficiency	Thin
Dampness, Phlegm	Rolling

Copyright © 2014 Anika Niambi Al-Shura. Published by Elsevier Inc. All rights reserved.

Western medicine:
- Consider whether the cause of the syncope may be due to hypotension. The pulse may also be slow and weak. Some elderly

patients with serious tachyarrhythmias may be monitored by a Western medicine practitioner who has prescribed medication to help maintain blood pressure at a low level for life preservation.

Auscultation:

- First palpate the apical impulse for left ventricular enlargement or hypertrophy.
- Listen for an early sign of hypertension, the fourth heart sound (S4) which indicates left atrium overwork.
- Listen for underlying sign of heart disease, the third heart sound (S3) which indicates left ventricular malfunction.
- Listen for a bruit in the carotid artery, indicating a possible murmur and arterial narrowing.

Genetic factors:

Genetic predispositions toward certain cardiovascular disease may be correlated with the causes of syncope in some patients.

1. miRNA
   Cardiovascular diseases include valvular disease and atrial fibrillation.
2. KCN (potassium channel) gene mutations and familial atrial fibrillation
   Three mutations on genes KCNE2, KCNJ2, and KCNQ1 involve the replacement of certain key amino acid proteins which make up the channels that regulate potassium flow in these diseases:
   - Romano−Ward syndrome/long QT syndrome
   - Short QT syndrome
   - Andersen−Tawil syndrome
   - Jervell and Lange-Nielsen syndrome
   - Long QT syndrome.

8.3 PART 3: CARDIAC CONCERNS

8.3.1 Arrhythmias

Arrhythmias contribute to poor cardiac output, which leads to hypotension and which is a main factor in episodes of syncope. However, patients often do not experience syncope during regular clinical visits. Frequent ECG monitoring may be necessary if the patient has been diagnosed with arrhythmias because of their significance. Arrhythmia is either tachycardia or bradycardia and is categorized in Chinese medicine as a deficiency state in xiong bi or chest obstruction, xin ji or palpitation, and xuan yun or vertigo.

15. What are the differentiations in Chinese medicine?
    tachycardia _____
    bradycardia _____
    Visit the course to find the answers.

8.3.2 ECG Findings

Prolonged EKG monitoring is suggested for patients with arrhythmia and for those without a cause for or history of syncope. However, it may have a few drawbacks such as an inconclusive correlation between the reported symptoms and the results. In addition, episodes of syncope rarely happen at the time of EKG monitoring, but evidence of rhythm disturbances, even if they are transient, can direct the health practitioner toward determining this as the cause.

8.3.3 Structural Problems

Structural problems can be detected during physical examination if a murmur is detected in auscultation, chamber or septum dimensions during echocardiography, heart arterial condition, and evidence of occlusion on cardiac catheterization. Other cardiopulmonary structural problems can include the following.

Structural problems	
Aorta	Aortic Stenosis and Aortic dissection
Cardiac tissue	Hypertrophic Cardiomyopathy and Acute Myocardial Infarction
Chamber	Cardiac Tamponade, and Left Artial Myxoma
Pulmonary	Pulmonic Stenosis and Pulmonary Hypertension

Copyright © 2014 Anika Niambi Al-Shura. Published by Elsevier Inc. All rights reserved.

8.4 PART 4: CEREBRAL CONCERNS

8.4.1 Cerebrovascular Disease

Syncope may be associated with vertebrobasilar disorders associated with ischemia such as vertigo, sensory, visual, and motor disorders. Patients with syncope connected with seizures notice an aura or changes in sounds, smells, and sensations before collapsing.

8.4.2 Imaging

An EEC and/or CT scan may be needed to rule out seizure from other neurological causes of syncope. An angiography may be needed to explore or monitor carotid artery disease and subclavian steal syndrome requires cerebral angiography.

NOTES

Log on at www.niambiwellness.com to access the companion course and quiz for Module 4.

Swellings and Fluid Build-Up

Section

Swellings and Fluid Build Up

CHAPTER 9

Clinical Significance of Edema

CHAPTER OBJECTIVES

After studying this chapter, you should be able to:

1. Explain the accumulation of edema according to the Starling equation.
2. Describe the renal factors in Western medicine and the Chinese medicine differentiation.
3. Describe the cardiopulmonary factors in Western medicine and the Chinese medicine differentiation.
4. Describe the hepatic factors in Western medicine and the Chinese medicine differentiation.

9.1 PART 1: DEFINITION

In Western medicine, edema is the accumulation of excess fluid in the interstitium. In Chinese medicine, edema is classified as shui zhong (shoo-ay jong).

9.2 PART 2: ETIOLOGY

The Starling equation, also known as the Frank–Starling heart law, was created in 1896 by Ernest Starling who was an English physicist.

$$J_v = K_f([P_c - P_i] - \sigma[\pi_c - \pi_i])$$

J_v is the rate of movement across the capillary membrane
K is the constant membrane permeability
P_c is the hydraulic pressure in the capillaries
P_i is the hydraulic pressure in the interstitium
π_c is the oncotic capillary pressure
π_i is the oncotic interstitial pressure.

Diagnosing in Cardiovascular Chinese Medicine. DOI: http://dx.doi.org/10.1016/B978-0-12-800121-9.00009-6
© 2014 Elsevier Inc. All rights reserved.

He used the equation to describe the Starling forces of hydrostatic and oncotic pressure of interstitial fluid movement across capillary membranes and lymphatic vessels. Oncotic pressure moves fluid out, while hydrostatic pressure moves the fluid in.

Failure which contributes to fluid accumulation leading to edema is caused by an increase in hydrostatic pressure especially concerning the retention of water and sodium in the kidneys, a reduction in oncotic pressure between the tissues and within blood vessels which causes increased permeability and obstruction within the lymphatic system causing a fluid stasis.

Two major pathological factors include glomerulonephritis and renal failure. Edema can be evident anywhere on or within the body. On the surface of the body, there are two types: pitting edema which leaves an impression in the skin when pressed with a finger and nonpitting leaves no impression on the skin when pressed.

In Chinese medicine, shui zhong is differentiated as lung wind invasion, spleen yang deficiency, spleen damp invasion, kidney yang deficiency, liver excess, and heart yang deficiency.

9.2.1 Renal Factors
Starling forces will sacrifice plasma volume with interstitial fluid accumulation, and the kidneys will retain salt and water contributing to nephrotic syndrome. Glomerulonephritis with hypoalbuminemia contributes to edema. Compensation causes increased arterial pressure which will eventually affect the lungs and heart, especially the left ventricle.

1. What are the Chinese medicine differentiations?

Visit the course to find the answers.

9.2.2 Cardiopulmonary Factors
Cardiopulmonary disorders involved in the formation of edema include the right and left heart failure. Right heart failure involves the elevation of right atrial pressure which affects the lungs and lymphatic flow. Left heart failure involves the left ventricle, myocardium, and systemic pressure is affected contributing to peripheral edema.

2. What are the Chinese medicine differentiations?

_____

Visit the course to find the answers.

9.2.3 Hepatic Factors

Acute liver inflammation and cirrhosis create a complex systemic situation that contributes to accumulation of ascites, which will eventually cause an extrinsic compression of the inferior vena cava. The hydraulic pressure within the capillaries will eventually contribute to edema in the extremities.

3. What are the Chinese medicine differentiations?

_____

Visit the course to find the answers.

NOTES

CHAPTER 10

Inquiry and Examination in Western Medicine and TCM

CHAPTER OBJECTIVES

After studying this chapter, you should be able to:

1. Explain Chinese medicine perspective of edema.
2. Discuss the significance of hereditary angioedema.
3. Describe the appearance of edema and physical examination.
4. Describe the inquiry process and determining the cause of edema.

In Chinese medicine, edema shui zhong (shoo-ay jong) is often differentiated as lung wind invasion, and nearly always differentiated as kidney yang deficiency, liver blood stagnation, and heart yang deficiency. It is a pathogenic condition which is not treated as an isolated condition and relies on Western medicine diagnostic methods to determine the magnitude of the edema in relationship to serious chronic diseases. There are at least three complications which are often involved in the development of isolated and peripheral forms of edema: congestive heart failure, nephrotic syndrome, and liver cirrhosis.

4. Peripheral edema indicates what cause and actions?

_____

Visit the course to find the answers.

10.1 PART 1: INQUIRY

When you ask the patient a question about the accumulation of edema, you should inquire about the awareness of body changes where edema is evident. Often the patient may notice that clothing, shoes, or rings fit tighter than usual or that the face, eyes, neck, and waistline may seem puffier at different times of the day. The persistence of the

Diagnosing in Cardiovascular Chinese Medicine. DOI: http://dx.doi.org/10.1016/B978-0-12-800121-9.00010-2
© 2014 Elsevier Inc. All rights reserved.

edema and whether the fluid accumulation is noticed at certain times of the day or is always present is important. Also edema is known to occur with premenstrual syndrome.

5. _____
   Visit the course to find the answers.

When you ask the patient about the time duration of these changes, consider the acute and chronic nature of the onset and development:

6. _____
7. _____
   Visit the course to find the answers.

When asking patients about other symptoms, consider what may point toward the heart, lung, or kidneys when the edema is apparent:

8. _____
9. _____
   Visit the course to find the answers.

10.2 PART 2: PHYSICAL EXAMINATIONS

These patients may need to be referred to a specialist who can monitor and provide additional diagnostic testing.

10.2.1 Patient Symptoms
10.2.1.1 Inspection and Palpation
10.2.1.1.1 Inspection of the Head and Face
Observe any tissue swelling located:

10. _____
11. _____
12. _____
13. _____
    Visit the course to find the answers.

This is evident in lung wind water syndrome in Chinese medicine and hereditary angioedema.

10.2.1.2 Palpation of the Bloated Abdomen for Ascites
Have the patient lie down face up. While standing and facing the right side, place the left hand on the left side of the waist. With the right

hand pat the right side of the abdomen. Ripples in the flesh may be indication of fluid accumulation. What are the differentiations and the disease?

14. _____
15. _____
16. _____

Visit the course to find the answers.

10.2.1.3 Palpation of the Limbs for Edema
• Acute edema will cause
  17. _____
• Chronic edema will
  18. _____

Visit the course to find the answers.

This is physical evidence of kidney yang deficiency, nephrotic syndrome and in some cases right or left heart failure.

10.2.1.4 Kussmaul's Sign
A rise in jugular venous pressure on physical examination which indicates right heart failure. This is evident in heart yang and qi deficiency and in congestive heart failure.

10.2.2 Symptoms of Edema

Symptoms of edema		
Cardiopulmonary	Increased abdominal circumference, dyspnea, facial swelling, edema, pulmonary congestion, ascites, lower leg edema and sacral edema	-Pleural effusion -Lung disease -Left ventricular failure -Right ventricular failure
Nephritic	Fluid accumulation in the lower legs with hyperpigmentation and peripheral edema	-Glomerulonephritis -Congestive heart failure -Nephritic syndrome
Hepatic	Ascities	-Liver cirrhosis
Lymphatic	Localized obstruction	-Lymphadenopathy -Lymphangiitis

Copyright © 2014 Anika Niambi Al-Shura. Published by Elsevier Inc. All rights reserved.

10.2.2.1 Location of Edema

Location of edema	
Peripheral (all four extremities)	Cardiac failure, renal failure
One arm	Lymphatic obstruction and lymphadenopathy and venous occlusion
Both arms	Superior vena cava occlusion
One leg	Lymphatic obstruction, deep vein thrombosis and phlebitis
Both legs	Chronic venous insufficiency, nephritic syndrome, glomerularnephritis and left and right ventricular failure

Copyright © 2014 Anika Niambi Al-Shura. Published by Elsevier Inc. All rights reserved.

Blood Pressure:

This is taken in sitting, standing, or supine positions and readings indicate stages of hypertension with edema. Fluid pressure in the vessels will raise the blood pressure. It will be necessary to notice the location of the edema to determine the possible cause.

Pulses:

Common Chinese pulse diagnosis:

Pulses	
Lung wind-water	Slippery and floating
Lung and kidney deficiency	Weak and deep
Heart yang (and qi) deficiency	Weak, knotted and deep
Dampness, phlegm	Rolling

Copyright © 2014 Anika Niambi Al-Shura. Published by Elsevier Inc. All rights reserved.

Western medicine:

- In Western medicine, consider the increase in pulse rate with the presence of edema.

Genetic Factors:

Genetic predispositions toward certain cardiovascular disease may be correlated with the causes of edema in some patients.

10.2.3 Hereditary Angioedema

This is an autosomal dominant condition which is due to deficiency of the C1 inhibitor protein of serpins. The symptoms of edema may be idiopathic or due to exposure to a pathogen. It is characterized by facial or tongue swelling and peripheral edema. Because of the idiopathic nature of hereditary angioedema in some patients it may be difficult to treat the edema which may accumulate with chronic pathology concerning the heart, lungs, liver, or kidneys.

19. What are the three types?

_____

_____

_____

Visit the course to find the answers.

NOTES

Log on at www.niambiwellness.com to access the companion course and quiz for Module 5.

Gait

CHAPTER 11

Clinical Significance of Claudication

CHAPTER OBJECTIVES

After studying this chapter, you should be able to:

1. Explain the definition of claudication.
2. Describe the Chinese medicine terminology for this condition.
3. Explain the cause of claudication.

11.1 PART 1: DEFINITION

Claudication is a painful, cramping sensation in the muscles due to movement or exercise. It can be experienced anywhere in the body, but the focus on the legs has an important clinical significance as a sign of cardiovascular disease. In Chinese medicine, claudication is considered in the category of vessel bi and blood stagnation.

1. What is important to realize about claudication during clinical evaluation?

   _____

   Visit the course to find the answers.

11.2 PART 2: ETIOLOGY

The clinical significance is that it may indicate the presence of atherosclerosis as a chronic condition in one or both legs. Claudication occurs when the blood flow through the vessel may be obstructed causing pain, weakness, or extreme fatigue. The severity of the symptoms is in direct relationship to the degree of the stenosis and the progression rate from the age of onset.

The age range of onset is between 40 and 70 years of age, more common in males than females, and the associated risk factors are due to disease and lifestyle choices such as diabetes mellitus, hypertension,

Diagnosing in Cardiovascular Chinese Medicine. DOI: http://dx.doi.org/10.1016/B978-0-12-800121-9.00011-4
© 2014 Elsevier Inc. All rights reserved.

cigarette smoking, and alcoholism. The painful awareness of the condition in the patient happens during movement or exercise. The resulting pain may be due to nerve stimulation in ischemic muscle tissue or blockages in the artery. The sensation is described as a dull aching pain which is exasperated by movement. If the patient attempts to continue movement, the pain will persist as a lingering tender area which prompts compensation and impairment.

2. What usually causes relief?

Visit the course to find the answers.

NOTES

CHAPTER 12

Inquiry and Examination of Claudication in Western Medicine and TCM

CHAPTER OBJECTIVES

After studying this chapter, you should be able to:

1. State the Chinese medicine classification of claudication.
2. Describe the purpose of the inquiry process.
3. Describe what to notice about gait and claudication during the physical examination process.

In Chinese medicine, claudication is classified under vessel bi and is merely a symptom within the greater syndrome to differentiate. In Western medicine, the physical examination should begin with observance of the patient's walking gait as they enter and exit the room and mount and dismount the examination table. The patient narrative is very important to report the factors involved with pain during walking and exercise.

Diseases which produce symptoms similar to claudication:

Buerger's disease	Muscle strain or tendonitis
Takayasu's arteritis	Peripheral neuropathy
Sciatica	Gout
Venous insufficiency	Arthritis
Nocturnal leg cramps	Peripheral artery disease
Myositis	Osteoporosis

Diagnosing in Cardiovascular Chinese Medicine. DOI: http://dx.doi.org/10.1016/B978-0-12-800121-9.00012-6
© 2014 Elsevier Inc. All rights reserved.

12.1 PART 1: INQUIRY

While observing walking difficulties during the patient clinical visit the first questions to ask are:

21. _____
22. _____
23. _____

     Visit the course to find the answers.

     Next ask about the details of the sensation, and the factors related to the need to rest.

- Would you describe the sensation as a dull cramping, aching, or burning?
- Would you describe the fatigue as tiredness or dead weight?
- Does the sensation occur only with movement or during rest as well?
- How long does the sensation last during movement?
- If the pain stops during rest, can you walk for the same distance or exercise for the same amount of time as before the sensation began?
- Which kind of movements cause the sensation the most?

     24. _____
     25. _____
     26. _____
     27. _____

     Visit the course to find the answers.

     Next ask the patient to demonstrate the movement which causes the sensation. A treadmill or a flight of stairs may be used to observe the patient movement. Allow the continuous movement until the claudication occurs. Take the patient's pulse after the sensation is reported to have occurred. If possible have a Doppler test done on the area.

12.2 PART 2: PHYSICAL EXAMINATIONS

Gait is the first observance in physical examination. This section should be done especially for those with other known cardiovascular disease symptoms and for the elderly patients.

Inspection and palpation
Gait

The trip from the waiting or triage room should provide some clue of the distance or amount of time the patient can travel before the occurrence of the sensation of claudication.

Mounting or dismounting the exam table

The muscle groups of the legs work together to help the patient mount the table. If there is pain in movement, the patient may create movements to compensate for the painful limb.

Blood pressure and heart rate

This is taken in sitting, standing, or supine positions and readings indicate stages of hypertension which is also associated with claudication in some patients. The increase in pulse rate is also apparent in the presence of claudication.

Genetic factors:

Genetic predispositions toward certain cardiovascular disease may be correlated with the predisposition of claudication in some patients.

Parental intermittent claudication

The children of parents with intermittent claudication may likely develop claudication and peripheral artery disease later during middle age.

Location of the claudication site

Location of claudication	Occluded artery
Greater trochanter	Iliac artery and profunda femoris
Thigh	Iliofemoral artery segment
Calf	Femoral artery
Foot	Tibial arteritides

Copyright © 2014 Anika Niambi Al-Shura. Published by Elsevier Inc. All rights reserved.

NOTES

Course Review Questions

1. Discuss the main sources of pain including the cardiovascular, pulmonary, and other causes of chest pain.
2. Describe the cardiovascular significance of dyspnea.
3. Describe the cardiovascular significance of orthopnea.
4. Describe the Chinese and Western medicine integrated differentiations for palpitations.
5. Describe the role of arrhythmias and structural problems in syncope.
6. Describe the inquiry process and how answers might distinguish the cardiac from the noncardiac causes.
7. Describe the appearance of edema and physical examination.
8. Describe the inquiry process and determining the cause of edema.
9. Describe the purpose of the inquiry process.
10. Describe what to notice about gait and claudication during the physical examination process.

Log on at www.niambiwellness.com to access the companion course and quiz for Module 6.

This also concludes the Integrative Anatomy and Patho-physiology in TCM Cardiology course. It is strongly suggested that you log onto the courses at the companion websites to review the course modules. Next, submit course documents and complete the final exam.

Upon passing the exam, you will receive completion certificates which include your name and practice license number, along with the specific number of credit hours awarded for this course. Electronic transmission of CEU and PDA credits will be sent to NCCAOM and your state medical board.

www.ingramcontent.com/pod-product-compliance
Lightning Source LLC
Chambersburg PA
CBHW072154020426
42334CB00018B/2005